WHAT·DOES·THE·CROW·KNOW?

Are lions capable of planning a hunt?

WHAT·DOES THE·CROW·KNOW?

The Mysteries of Animal Intelligence

BY MARGERY FACKLAM

ILLUSTRATIONS BY PAMELA JOHNSON

SIERRA CLUB BOOKS FOR CHILDREN
San Francisco

The Sierra Club, founded in 1892 by John Muir, has devoted itself to the study and protection of the earth's scenic and ecological resources — mountains, wetlands, woodlands, wild shores and rivers, deserts and plains. The publishing program of the Sierra Club offers books to the public as a nonprofit educational service in the hope that they may enlarge the public's understanding of the Club's basic concerns. The Sierra Club has some sixty chapters in the United States and in Canada. For information about how you may participate in its programs to preserve wilderness and the quality of life, please address inquiries to Sierra Club, 730 Polk Street, San Francisco, CA 94109.

A LUCAS • EVANS BOOK

Text copyright © 1994 by Margery Facklam
Illustrations copyright © 1994 by Pamela Johnson

First Edition

ACKNOWLEDGMENTS

Special thanks to Dr. Sally Boysen at the Primate Cognition Center at Ohio State University; Dr. Irene Pepperberg at the University of Arizona; poet Desi Vail and her guide dog, Rainy; and Jennifer Owings Dewey, who also wonders what the crow knows.

Library of Congress Cataloging-in-Publication Data

Facklam, Margery.
    What does the crow know? : the mysteries of animal intelligence /by Margery Facklam ; illustrations by Pamela Johnson.
        p. cm.
    Includes index.
    Summary: Raises the issue of whether or not animals are capable of thought, learning, remembering, and creativity, with examples of animal behavior that appears to be truly intelligent.
    ISBN 0-87156-544-7
    1. Animal intelligence — Juvenile literature.   [1. Animal intelligence.]   I. Johnson, Pamela, ill.   II. Title.
QL785.F33 1994
156'.3 — dc20   93–17811

Book and jacket design by Jean Krulis

This book is for

BRIAN THOMAS FACKLAM

from his grandma

.

CONTENTS

1   CLEVER HANS AND BORED CATS   9

2   AMAZING ALEX AND OTHER BIRD BRAINS   15

3   AN ELEPHANT NEVER FORGETS   22

4   COUNT ON IT   26

5   LEARNING TO DISOBEY   33

6   COMING UP WITH SOMETHING NEW   39

INDEX   46

1
CLEVER HANS
AND BORED CATS

Clever Hans taps out an answer

"How much is four plus five?" a man in the audience shouted at the horse.

What a question to ask a horse! But his owner, Wilhelm von Osten, nodded to Hans, a signal to "go ahead." In the silent room, the only sound was the tapping of the horse's hoof — nine times.

Amazing, people said. What a smart horse!

The horse could also multiply, divide, and subtract. With a nod or a shake of his head, he could answer yes or no to questions about music, history, and other subjects. People began to call the horse Clever Hans. It wasn't long before scientists came to Germany from all over Europe to see him, and most of them agreed that this smart horse really knew the answers.

To prove that Hans was really clever, von Osten asked a horse trainer, a circus manager, two zoologists, and a psychologist to watch Hans at work. After the performance, these experts agreed that Hans was truly amazing.

But one scientist wasn't convinced. Oskar Pfungst, a young psychologist, decided to try a test of his own. He wanted to see what Hans would do if a question was asked by a person who didn't know the correct answer. Pfungst discovered that Hans didn't know math or music or other subjects at all. He didn't understand the questions. But in a way, the horse really *was* clever, because he had learned to pick up cues, or signals, that even his trainer didn't realize he was sending. If von Osten — or any other questioner — so much as raised an eyebrow, changed his or her breathing, shifted a foot, or moved a shoulder, it was message enough to tell Hans which way to move his head or when to stop tapping.

Ever since von Osten's horse fooled scientists more than a hundred years ago, anyone who tries to test an animal's intelligence worries about making the Clever Hans mistake. Scientists are careful to set up "blind" tests in which the animal cannot get any clues from the person giving the test. It's not easy to find the right way to test an animal. It's hard enough to find the right way to test how smart a human is.

One test designed for people measures an individual's I.Q., or *intelligence quotient*. The score is a number that tells how well a person answers questions or figures out problems in comparison to other people of the same age who take the test. But people are smart in different ways. Some are good at math, while others have better language skills. Some are good mechanics, and others are all thumbs at figuring out how things work.

Background and culture, and where a person lives, can also affect test scores. A boy living in a big city might not be able to find edible plants or track an animal through a rain forest, any more than a boy raised in a rain forest would know how to program a VCR or use a computer. Even though both boys could be equally intelligent, it's unlikely that either one could pass a test based on skills learned in the other's environment. So it is with animals. Each animal is smart in a different way.

Two scientists made up a sixteen-part intelligence test that was supposed to measure a cat's ability to solve problems. It looked at how quickly a cat could perform various tasks, such as getting out from under a paper bag placed over its

A cat takes the paper bag test

11

Koko points to something good to eat

head, getting a piece of sticky tape off its nose, and getting at food wrapped in a napkin.

The test didn't work very well because the cats got bored with it. One cat loved hiding in the paper bag and didn't even try to get out. Another cat curled up and went to sleep during the test, and another hid under the couch. "If it feels as though you're pushing it, even the brightest cat will sit down and stare into space," said one trainer. "It's not stupid. It's just being a cat."

Dr. Penny Patterson is a scientist well known for her work with gorillas. After she taught a gorilla named Koko how to communicate in sign language, she wrote, "Koko gets bored with tests, and she'll do anything to change the subject."

Koko seems to feel that if she does badly enough on a test, Dr. Patterson will stop giving it to her. One of her favorite ways to fail is to stare at the right answer while she points to the wrong one.

Answers that seem right to an animal are sometimes scored as errors on tests made for people. On one kind of intelligence test, a young child is asked to point to things that are good to eat. The choices are a block, an apple, a shoe, a flower, and an ice cream sundae. When Koko was given the test, she picked the apple and the flower, which are right answers for a gorilla. But the flower is a "wrong" answer for this test.

Another question asks children to point out where they would run for shelter from the rain. The choices are a hat, a spoon, a tree, and a house. As any sensible gorilla would, Koko chose the tree. But the test wasn't made for sensible gorillas.

Intelligence is a combination of many things — thinking, learning, planning, remembering, and making decisions. Some learning comes easily to animals, while other things are impossible for them to learn. That's true of us as well. We could no more hunt like a pod of killer whales than killer whales could write. A dog can't read, but we can't follow a trail with our noses. Each animal uses different senses to learn what it must learn in order to survive in its habitat. Some animals have a sharp sense of hearing; others have eyes that see long distances or that adjust to darkness better than those of other creatures. Animals live in many different environments — underground, in the air, underwater, in deserts, in rain forests. All these differences make it almost impossible to compare intelligence among animals.

It's not easy to understand *thought* because we can't see it or feel it. We can't know what's going on in the mind of an animal any more than we can know what another human is thinking.

What does a lion think as it stalks a wary gazelle? What does a chimpanzee think as it pokes a stick into a termite's nest? Do these animals think about anything?

Dr. Donald Griffin is famous for his studies on animal intelligence. He wrote, "It makes sense that animals can think. If an animal thinks about what it might do, even in very simple terms, it can choose the action that will help it avoid dangerous mistakes. Thinking means survival."

But even if Dr. Griffin is right, and animals do think, can they remember things that happened in the past? Can they learn, and plan ahead, and teach others in their group? Are they creative? Do they have ideas? Are they aware of what they are doing? Or are they only programmed by the built-in patterns of behavior called instinct?

Some scientists look for answers in the laboratory. Others keep watch in the wild.

2
AMAZING ALEX
AND OTHER BIRD BRAINS

Alex takes his place at his work station

"Hello. I love you," Alex squawked when Dr. Irene Pepperberg walked into her laboratory at the University of Arizona. Alex is an African grey parrot. He was pacing up and down in front of his open cage. Behind him stood a cart piled with brightly colored keys, blocks, letters, small plastic cars, and other toys.

"I love you, too, Alex," said Dr. Pep-

Alex can answer the question "How many?" as long as there are no more than eight objects. He can also recognize shapes — round, triangle, and square, although he knows a square as a "four-corner." He was shown a tray full of objects — keys, a plastic triangle, a rock, a small green car, a red circle, several four-corner things, and several blue objects he had not seen before.

"What matter four-corner blue?" Dr. Pepperberg asked. Alex took his time. He tilted his head, the better to stare at each object with one shiny, dark eye. He poked some things with his beak and picked up others. Finally he chose the blue wooden block and answered the question: "Wood."

Alex has not just memorized these things. In order to make such choices, Alex must first understand the question. Then he has to compare the objects and find the right word for his answer. "It is original, logical thinking," Dr. Pepperberg said. She was surprised to find that Alex scores slightly better at answering questions about colors, shapes, and sizes when he is shown objects he has not seen before. Does he concentrate more when he sees something new, as we might?

Dr. Pepperberg knows about the Clever Hans problem, so she is careful not to give Alex any clues. When she is keeping score of the parrot's answers, she sits where she cannot see the object another team member is showing to Alex.

Once in a while Alex doesn't feel like answering questions. He turns his back or squawks, "No." Then he gets a break. Or he may purposely give the wrong answer over and over again. Sometimes he asks for a treat, such as raw pasta, which is one of his favorites.

Dr. Pepperberg believes that parrots are on the same level of intelligence as chimpanzees and dolphins. "No one knows how much more a parrot might know in the wild," she said. "But it's hard to study them in their natural habitat."

Alo and Kyaaro are just beginning their training

She has begun to teach two younger parrots, Alo and Kyaaro, and she is eager to see what differences there will be in what the three birds can learn and understand.

Parrots aren't the only birds that show original, logical thinking. Crows and ravens have always been famous for the surprising ways they solve problems.

One winter day, a scientist was driving through the Olympic Mountains in Washington. When he stopped at the roadside to eat his lunch, he tossed some crackers to a raven, which is the biggest member of the crow family. Even after the raven had eaten its fill, it kept on picking up crackers, probably to add to its *cache,* or collection of food. But the bird had a problem. To take one cracker at a time back to its hiding place meant leaving the rest of the crackers for blue jays and squirrels. Yet the raven couldn't pick up a second cracker without dropping the first one. What to do? The raven finally solved the problem by sticking one cracker at a time edgewise into a snowbank, like slices of bread in a toaster. In a few minutes, the bird had gathered six or seven crackers close together in a neat row. Then the raven opened its big beak, snapped up all the crackers at once, and flew off to its hidden cache.

A raven arranges crackers for easy pickup

Another true story about the crow family comes from Finland. Some crows were watching two ice fishermen haul in their lines and put fresh bait on their hooks. After the men dropped the lines back into the holes in the ice, they set a flag at each hole that would flip up to signal when a fish had taken the bait. The men had just settled into their warm shed to wait when the first flag snapped up. But before they could get to it, one of the crows flew down, grabbed the line in its beak, and backed away from the hole. When the crow dropped the line, of course the line slipped back down the hole. Again the crow picked up the line and walked away with it. But this time, when the crow dropped the line, it was careful to keep both feet on it. The crow walked on the line all the way back to the hole, where it picked it up again and backed away. The bird did this over and over, until it had hauled in the line and a fish flopped onto the ice.

If you watch crows feeding in a cornfield, chances are you'll see one crow perched on a fence post. That's the lookout. Of course, not everyone agrees that this crow is really standing guard. Some think it's only a crow that's not hungry. But if two people walk into the cornfield, that guard crow caws a signal, and a few nearby crows fly into the trees. If those people are carrying guns over their shoulders, the lookout crow's call is quite different. It's a harsher, longer cry that seems to warn the other crows of danger, and the whole flock takes off in a flurry of flapping wings and raucous caws.

But what does the crow really know? It certainly doesn't take crows long to figure out that a scarecrow is harmless, but can they understand that a human with a long stick over the shoulder is more dangerous than a human without one? The author Henry Ward Beecher once wrote, "If men had wings and bore black feathers, few of them would be clever enough to be crows."

3
AN ELEPHANT
NEVER FORGETS

An elephant muffles its bell with mud

The author J. H. Williams, who called himself "Elephant Bill," spent twenty years with elephants in the forests of Burma. He wrote, "An elephant never stops learning be- cause he is always thinking." He had often seen tame work elephants stuff mud into the bells around their necks before they sneaked out at night to steal bananas. But did one smart elephant figure out how to silence a bell and teach the others, or did each elephant discover this trick for itself?

As a general rule, the larger the animal and the longer it lives, the more it thinks, learns, and teaches it young. The huge elephant, with its life span of more than sixty years, is a perfect example of that rule.

Elephants that work in logging camps easily learn commands such as "Stop," "Go," and "Back up." They also know the difference between commands to "Push the object with your feet" and "Push the object with your head." After a few years on the job, elephants seem to know what they are expected to do, and they work with little direction from humans.

For thousands of years, people have told stories about the smart things elephants can do. But stories aren't enough for scientists. An idea isn't a fact until it's proven. An experiment done by one scientist must show the same results when it is repeated by other scientists. Some of the oldest stories about elephants tell about their long memories. But is it true that an elephant never forgets?

Some scientists who were curious about an elephant's memory tested a female elephant at a zoo in Germany. First they taught her some short melodies. After she knew them well, the melodies were changed. They were played faster, then slower, with different timing and in different rhythms. They were played on different instruments, in higher tones, and in lower tones. But through all the changes, this elephant recognized every melody she had learned.

In another series of tests, the same elephant learned to match patterns on pairs of cards. After 600 trials, the people giving the tests were getting weary, but the elephant wasn't bored at all. In fact, toward the end of the tests, the elephant's scores got better. A year later, the elephant still remembered the melodies and the patterns on the cards.

Killer whales surround a school of fish

Dr. Donald Griffin believes that animals not only learn and remember but can also plan ahead. Lions are predators, and like all predators their instincts drive them to hunt. But when Dr. Griffin watched four female lions hunt down a wildebeest, he was convinced that more than instinct was at work. Two of the lions climbed a small mound and sat where a herd of wildebeests could see them. A third lion, hidden by tall grass, crawled on her belly into a ditch that ran between the two watching lions and the wildebeest herd. For a few minutes all was quiet. Then suddenly the fourth lion charged out of a small woods behind the herd. Startled, the wildebeests thundered toward the ditch. As they bounded over the ditch, the lion hiding there leaped out and pounced on a wildebeest. She killed it quickly, and all four lions joined in the feast. Some who study animal behavior say that the lions just hap-pened to be in those positions, but Dr. Griffin believes they had learned to work together. "It seems reasonable," he wrote, "that lions are capable of plan-ning their hunting behavior."

Killer whales and dolphins also hunt in groups. When a pod of killer whales surrounds a large school of fish, they swim in smaller and smaller circles, as if closing a net. Dolphins hunting for food will swim in formation, like soldiers marching shoulder to shoulder. When their sonar signals bounce back from a school of fish, the dolphins surround their prey. Some of the dolphins keep circling the fish while others eat; then they change places.

Many other animals, including wolves and chimpanzees, are coopera-tive hunters, too. Cooperation and plan-ning help many species of big-brained animals survive where there is a lot of competition for food.

4
COUNT ON IT

Darrell studies fractions

D arrell is working on fractions. He's twelve years old, and so far he knows only such simple fractions as one-half and one-fourth. But Darrell is a chimpanzee!

Who would have guessed that a chimpanzee could understand fractions? Dr. Sally Boysen did. She runs the Primate Cognition Center at Ohio State University, where Darrell lives with four other chimpanzees — Kermit, Sarah, Sheba, and Bobby.

Dr. Boysen shows Darrell half a pear, and Darrell points to a number card showing ½. "Darrell has to understand that there are whole things and parts of things, and that these numbers of things have names," Dr. Boysen says. But if Darrell is shown a fourth of a pear, he does not yet seem to understand that it is one-fourth of a whole pear. He still needs to watch as the object is being divided.

Not all chimpanzees are equally good at numbers. As with humans, each animal has different abilities. Kermit is the same age as Darrell, but he hasn't caught on to counting yet. Eleven-year-old Sheba learned to count when she was only five. She had already learned colors and could easily point out things that were the same or different. At first she counted gumdrops, "tagging" (touching) each one the way a young child counts. Then she learned to recognize the numbers that represented these amounts.

One day when they were working outdoors, Sheba watched Dr. Boysen drop three peaches into one box and three more peaches into another box a few feet away. Together they walked back to a display of number cards. "How many peaches in the boxes?" Dr. Boysen asked. Sheba points to the six.

"I couldn't have been more surprised," Dr. Boysen said. Sheba understands the numbers zero through eight, as well as the numerals and English words for them. She knows that some numbers are larger or smaller than others, and she can put them in order. But Sheba had never been taught to add objects or numbers. It meant that Sheba had to count the first group of peaches, hold that number in her head, and continue counting when she saw the second group of peaches. When Dr. Boysen substituted number cards for peaches, Sheba added those just as well. It was once thought that only humans could understand at this level.

Dr. Boysen has invented a harder game for Sheba. In this one, she puts two groups of candies on a table in front of Sarah's cage. (Sarah's only job in this test is to eat candy.) Dr. Boysen asks Sheba, "Which candies do you want to give to Sarah?" Sheba knows she will get the ones that are left.

But Sheba points to the largest group of candies first. "You want to give Sarah five?" Dr. Boysen asks. "Too bad. Sheba gets only one."

So far in these tests, Sheba always points to the largest number of candies first. But when number cards are used in place of the candies, that changes. Then Sheba points to the smaller number first, so that Sarah gets fewer candies and Sheba gets more. But why would Sheba choose to give more candy to the other chimp, even though she understands the numbers? Dr. Boysen thinks that maybe Sheba is so eager to get the candy that she points to the larger number of candies before she thinks of the consequences. But when Sheba sees the number card instead of the candy, she understands that if she chooses the smaller number to give away, she can keep the larger number for herself.

The chimpanzees love to go into the computer room. They all want a turn, and they push and shove at the doorway until Dr. Boysen says in a stern voice, "No, Darrell, it's Sheba's turn. Kermit, you can have a turn later."

The chimpanzees' computer room is separated from Dr. Boysen's room by a wall with a thick glass window. Sheba is so excited and eager to get started that she bangs on the glass. "Just a minute, Sheba," Dr. Boysen says. "The window's dirty." She hands a wet paper towel to Sheba, who scrubs at the smudges, stopping to lick one or two sticky spots. After she wipes her computer screen, she's ready to work.

Dr. Boysen sits at her computer on the other side of the window, where she can see Sheba but not Sheba's computer

Sheba takes a turn at the computer

screen. It is a "blind" test. Dr. Boysen sets three empty spools on the windowsill. "How many?" she asks. In a flash, Sheba touches the number 3 on her screen, and her answer shows up on Dr. Boysen's screen. "Good girl. Do you want Gummy Bears or M&M's?" Sheba points to the jar of Gummy Bears, and Dr. Boysen sends three of the candies down the tube to Sheba's side of the window.

The numbers on Sheba's screen change places with each question. The number 3 may be in the top right corner first, but next time it might show up at the bottom left of the screen. Sheba has to know that the number 3 represents three objects. "Sheba understands that anything can be counted," Dr. Boysen says.

Five-year-old Bobby doesn't understand numbers yet. He is learning the basics, such as shapes and colors. Mostly he likes to fool around and play. Dr. Boysen says, "We spend fifteen minutes in school and two hours at recess."

Dr. Boysen wants people to know that chimpanzees are not justy hairy humans. She smiles and says, "But close."

Chimpanzees are, indeed, close cousins of ours. About 98 percent of our DNA is exactly like a chimpanzee's. *DNA (deoxyribonucleic acid)* is the substance that makes up the genes in each cell of our bodies. It carries the blueprints that make us what we are. But the 2 percent that is different makes for a big difference. For one thing, it is what makes humans the only animals with a written language.

Dr. Boysen admits that studying animals in captivity is quite different from watching them in the wild. But it's the only way to get really close to them. "I just don't want people reading things into these studies or exaggerating what we find," she said. "There's a lot to learn about these animals."

What Dr. Boysen learns about chimpanzees may encourage scientists to study other animals that have a sense of numbers. Eskimo hunters say that wolves can count up to seven, but no higher. When seven hunters follow a pack of wolves, the wolves become disturbed if one of the hunters drops out of line to circle around the animals. But if there are more than seven hunters, the wolves seem not to notice their comings and goings, as long as they can always see seven of them.

Crows and ravens can keep track of numbers, too. Jennifer Owings Dewey wrote a book about an orphaned raven that she had rescued and named Clem. Like other ravens, Clem collected things. In the wild, ravens and crows keep a cache of food to help them survive the winter. But Clem had plenty of food, so he saved almost anything he could carry — jar lids, string, buttons, stones, even socks. One day Jennifer found Clem's cache of stones under a tree. She put the stones in an open box, and then

Clem's unusual cache included jar lids and buttons

just to see what Clem would do, she added three extra stones. A few days later, the three extra stones were gone. Jennifer took away three more stones. When she checked again, she found that Clem had added three more stones to replace the ones she had taken.

Could Clem really keep track of numbers? Jennifer kept experimenting. She made several rows of stones — a row of three, a row of five, a row of four. After she was sure that Clem has seen these rows, she took away a stone from the row of four. Clem added one stone. If she added two more stones to the row of three, Clem would take away two stones. He seemed to remember the number of stones that were in each row when he first saw them, because he always removed or replaced the correct number to keep the rows in order. Most ravens and crows can keep track of numbers up to six, and some up to eight. Alex, the African grey parrot, can also name quantities up to eight.

This ability to keep track of numbers of things must help animals survive. But why do wolves keep track of up to seven things? Why do crows and ravens keep track of up to six or eight? Why do we? When people glance at groups of objects without enough time to count them, they score no better than ravens or wolves.

In 1887, an English scientist named Joseph Jacobs tested short-term memory in children. He found that a human's *short-term memory* — the part of the brain that remembers things for just a little while — seems to work better when it is "chunked." Like many other animals, we, too, tend to remember things in chunks of six, seven, or eight.

5
LEARNING TO DISOBEY

A Seeing Eye dog guides its blind owner

1 magine that you are blind. You are standing at a curb, waiting to cross a busy street. You can't see the traffic. Noise surrounds you. Horns honk, motors rumble, people shout, heels click on the sidewalk, and a radio blasts rock music. Your left hand is on the handle of a harness worn by a German shepherd, and you have to trust that dog to take you to the other side of

worked outdoors in Texas had to learn the warning sound made by a rattlesnake. When the dog heard a rattler, it did not go toward the snake, but only turned its head in the direction of the sound. That was enough to tell the man he should stay away from a particular clump of grass or pile of rocks.

Many deaf people can drive safely as long as a companion dog sits in the front seat, ready to put a paw on the driver's knee when it hears a siren, honking horn, or screeching tires.

Friendly, easy-going Labrador retrievers are especially good helpers for people in wheelchairs. A group called Canine Companions for Independence gives eight-week-old puppies chosen for this work to families to raise for fourteen months. The dogs have to learn forty-nine commands before they go to advanced training, where they learn to pull a wheelchair, open doors, turns lights on and off, and pick up dropped objects. By the time a dog meets its human partner, it knows eighty-nine commands. Then they both go to a two-week "boot camp," where they learn more about working together.

Now imagine that you are sitting in a wheelchair and cannot move your arms or legs. You want to turn on the TV or scratch your nose or get a drink, but you can't. You just have to wait for help. Even a companion dog can't help you, but a monkey can.

An organization called Helping Hands trains capuchin monkeys to live with people who are paralyzed. The capuchins learn to fetch things that the person points to with the bright light of a laser pointer, which is mounted on the wheelchair and held in the person's mouth. Instead of expecting a monkey to respond to commands such as "Please bring me that magazine on the floor," the person simply points to the magazine. A monkey can turn the pages of a book on a reading stand, or get a wrapped sandwich and unwrap it. The

A capuchin monkey fetches a drink for its owner

ing that it's annoyed. Immediately, Karen blew her whistle and tossed Malia a piece of fish.

"That was enough for Malia," Karen said. "She got the message. She slapped, ate her fish, slapped, ate, slapped. In less than three minutes she was motorboating around the tank, pounding her tail on the water."

Karen Pryor was showing the audience *operant conditioning,* in which the animal is the operator — the one who starts the action. The first step is to teach the animal that the whistle is a signal that means a reward is coming. Trainers call the whistle a *bridge* because it bridges the gap between the action and the reward. It doesn't take long before the animal learns to respond to the whistle alone, knowing that the reward will come later.

Using the whistle to change the animal's actions a little at a time is called *shaping.* By blowing the whistle every time a dolphin turns to the right, for ex-

ample, a trainer can have the animal swimming in a tight circle in just a few minutes.

At the first show on the day after Malia had motorboated around the tank, she slapped her tail, but nobody tossed her a fish. What now? she seemed to be saying as she grew more annoyed. In this case, Karen was asking Malia to "invent" a new action by not rewarding her for doing something she had shown the audience before. Finally Malia grew so annoyed that she threw her whole body into the air and came down sideways, hitting the water with a huge splat. Instantly, Karen blew her whistle and tossed a piece of fish. It was clear that Malia got the message, because she splatted on her side again and again.

For a few days, the trainers waited at each show for Malia to do something different. There were a couple of embarrassing shows in which Malia went splashing around the pool doing every trick she knew, but nothing new.

A whistle and a fish reward Malia's creativity

Then one day, during the last show, Malia swam out of her holding tank and circled the pool, waiting for a cue, but there was no whistle. "Instead of going through all the old routines," Karen said, "she suddenly got up a good head of steam, rolled over on her back, stuck her tail in the air, and coasted about fifteen feet with her tail out, as though she were saying, 'Look, Ma, no hands.'"

After that, Malia did amazing things, show after show. She spun in the air, swam upside-down, and revolved like a corkscrew. "She thought of things we could never have imagined," said Karen. "Sometimes she was very excited when she saw us coming in the mornings. I had the unscientific feeling that she sat in her holding tank all night, thinking up stuff, and rushed into the first show with an air of 'Wait until you see this one.' Originality — rare, but real, in animals. Almost never observable in a laboratory situation."

A lot of people believe that animals go through life unaware of what they are doing. Many animals probably do. We humans certainly do a lot of things automatically, without thinking. But there is good evidence that some animals are not only aware of what they are doing, but, like Malia, can be creative as well.

At one time, we thought that humans were the only animal that invented and used tools. But now there is a long list of animal tool-users. Chimpanzees make special sticks as tools for digging out termites, and they crunch up leaves to sponge up fresh drinking water. Sea otters collect and use stones to open abalone shells. Darwin finches use thorns to pry insects out of tree bark. Egyptian vultures throw stones to break open ostrich eggs. Recently, a scientist saw a group of dolphins, and each one had a sea sponge stuck on the end of its beak. Were those

A sea otter uses a stone to open an abalone shell

sponges a kind of tool the dolphins used to collect small bits of food, or were they part of a game? Nobody knows, but scientists are certainly watching those dolphins.

Before we knew about dolphins like Malia, we thought that only humans had original ideas. And before elephants started to paint, we thought that only humans were artists.

In the wild, elephants lead busy lives, roaming hundreds of miles in their home territories. But captivity can be boring. In a zoo, there's little to do. Siri is an Asian elephant who has lived at the Burnett Park Zoo in Syracuse, New York, since 1970, when she was two years old. Nobody knows when she began to draw, but in 1976, one of her keepers noticed designs on the floor of her pen. Siri had been making marks with a pebble that she held in her trunk. A few years later, another keeper gave Siri a big carpenter's pencil. At first Siri looked it over, tasted it, and scratched her back with it.

But when the caretaker put the pencil point on a huge sketch pad, Siri moved her trunk and the pencil left a mark. After that, Siri filled page after page with arcs and lines that one art professor said were very much like the abstract paintings of human artists. In the next two years, Siri did more than 200 drawings with crayons, colored pencils, or brushes and paints.

When the curator at the Washington Park Zoo in Portland, Oregon, heard about Siri, he didn't think she was so amazing. "All our elephants draw," he told a reporter.

But of all the elephant artists in captivity, Ruby is probably the most famous. She lives at the Phoenix Zoo in Arizona. She was only seven months old when she was separated from her mother in Thailand. Even though she had human friends, she wasn't content. She charged at trucks and stomped on ducks and geese that got too close. Her keepers could never be quite sure when Ruby

Ruby puts the finishing touches on a painting

might be a danger to them. But since she has learned to paint, Ruby has been a much happier elephant, and she hasn't stomped a single duck.

When someone brings out her paint supplies, Ruby taps her trunk on the color she wants to have put on her brush. She always waits for the color she has requested before taking the brush. After she has finished a painting, which usually takes about ten minutes, she shuffles a few steps away from the easel, flips up her trunk, and often pats herself on the cheek. She doesn't like it at all if anyone changes her painting!

Does Ruby know what she's doing? The zoo people think so, because Ruby's paintings often reflect what she sees. Her most famous painting is called *Fire Truck*. After she had watched a fire engine arrive at the zoo with its blue light flashing, Ruby covered her canvas with bold red strokes, topped with a bright blue splotch.

Dr. Sally Boysen says, "All our chimpanzees love to paint." When I met Bobby, the youngest chimpanzee in her care, the first thing he did was grab my notebook and pen. He hunkered down in a corner of his playroom, with one arm curved around the paper like a kid who doesn't want anyone to see his work, and he doodled circles, arcs, and lines on every page.

Which animal is the smartest? Nobody knows, because there is no way to compare them. Each animal is as smart as it needs to be, or it wouldn't survive. Most scientists who study animals tend to agree with Dr. Roger Payne, who has studied whales for many years. He says that the longer he studies whales, the more there is to know. And that's about it. One question answered opens the door to more questions.

The human animal may be able to write a book, perform brain surgery, or build a computer. But those skills don't make us better than other animals. They only make us different.

INDEX

African grey parrots, 15–19, 32
Alex (parrot), 15–19, 32
Alo (parrot), 19
Artists, animal, 43–45

Beecher, Henry Ward, 21
Birds. *See* Crows, Parrots, Ravens
Blind, companion dogs for the, 33–35
Blind tests, 10, 29
Bobby (chimpanzee), 27, 29–30, 45
Boysen, Dr. Sally, 26–30, 45
Bridge, in training, 40
Burnett Park Zoo, 43

Cache, 19, 30
Canine Companions for Independence, 36
Capuchin monkeys, 36–38
Cats, 11, 13
Chimpanzees, 14, 18, 25, 26–30, 42, 45

Clem (crow), 30–32
Clever Hans (horse), 9–10
Companion dogs, 33–36
Computers, and chimpanzees, 28–29
Counting
    chimpanzees, 27–29
    crows and ravens, 30–32
    parrots, 18, 32
    wolves, 30, 32
Creativity, 39–45
Crows, 19–21, 30, 32

Darrell (chimpanzee), 26–27, 28
Darwin finches, 42
Deaf, companion dogs for the, 35–36
Dewey, Jennifer Owings, 30, 32
DNA (deoxyribonucleic acid), 30
Dogs, 14, 33–36
Dolphins, 18, 25, 39–41, 42–43

Egyptian vultures, 42
"Elephant Bill" (J. H. Williams), 22
Elephants, 22–23, 43–45

Finches, Darwin, 42

Gorillas, 12–13
Griffin, Dr. Donald, 14, 25
Guide dogs, 33–36
Guiness Book of World Records, 16

Helping Hands, 36, 38
Henry VIII, King, 16
Horses, 9–10
Hunting, 25

Imitation, parrots, 16–17
Instinct, 14
Intelligence tests, 10–13
Intelligent disobedience, 34, 35, 38
I.Q. (intelligence quotient), 10–11

Jacobs, Joseph, 32

Kermit (chimpanzee), 27, 28
Killer whales, 14, 25
Koko (gorilla), 12–13
Kyaaro (parrot), 19

Lions, 14, 24–25
Lookout, crow as, 21

Malia (dolphin), 39–42, 43
Memory, elephants, 23
Monkeys, 36–38

Number use
    chimpanzees, 26–30
    crows and ravens, 30–32
    parrots, 18, 32
    wolves, 30, 32

Operant conditioning, 40
Originality, 39–45
Original thinking, birds, 18–21
Osten, Wilhelm von, 9–10

Painting
    chimpanzees, 45
    elephants, 43–45
Parrots, African grey, 15–19, 32
Patterson, Dr. Penny, 13
Payne, Dr. Roger, 45
Pepperberg, Dr. Irene, 15–19
Pfungst, Oskar, 10
Phoenix Zoo, 43
Primate Cognition Center, 26
Pryor, Karen, 39–41

Rainy (dog), 35
Ravens, 19, 30–32
Ruby (elephant), 43, 45

Sarah (chimpanzee), 27, 28
Sea Life Park, 39
Sea otters, 42
Seeing Eye, Inc., 34
Shaping, 40
Sheba (chimpanzee), 27–29
Short term memory, 32
Sign language, 13

Siri (elephant), 43
Sparkie Williams (parrot), 16
Speaking, parrots, 15–18

Tool use, 42–43
Tricks, dolphins inventing, 40–42

Vail, Desi, 34–35

Vultures, Egyptian, 42

Washington Park Zoo, 43
Whales, 45
    killer, 14, 25
Wheelchair users, companion animals for, 36–38
Williams, J. H. ("Elephant Bill"), 22
Wolves, 25, 30, 32